MIND

MIND

AN EMERGENT PROPERTY

Simeon Locke

Copyright © 2014 by Simeon Locke.

ISBN: Softcover   978-1-4990-1414-3
           eBook   978-1-4990-1415-0

All rights reserved. No part of this book may be reproduced or transmitted in any form or by any means, electronic or mechanical, including photocopying, recording, or by any information storage and retrieval system, without permission in writing from the copyright owner.

Any people depicted in stock imagery provided by Thinkstock are models, and such images are being used for illustrative purposes only.
Certain stock imagery © Thinkstock.

This book was printed in the United States of America.

Rev. date: 05/21/2014

To order additional copies of this book, contact:
Xlibris LLC
1-888-795-4274
www.Xlibris.com
Orders@Xlibris.com
618525

Contents

I. Emergent Properties 7

II. Emergence Of The Psychophysical 12

III. Organization Of The Nervous System 15

IV. Evolution 21

V. Environment 26

VI. Percepts And Concepts 29

VII. Language 33

VIII. A Model Of Reality 36

IX. Mind 41

X. Conclusion 46

Chapter I

Emergent Properties

The creation of emergent properties seems to be a basic principle of nature. It appears in animate and inanimate situations, in chemical, physical and biological systems, in individuals and communities and in tangible and intangible form. The properties emerge from an aggregate of modules; as few as two (NaCl yields salt), to a large collection. The modules may be identical or dissimilar. For a given property a critical number of modules is required. Below that number the effect may be additive. When the critical element is added a new property emerges. It does not inhere to any of the modules. It is neither predictable from nor reducible to the components. The simplest example is water, a liquid composed of two gases. It's characteristics-wetness, for example-are not present in it's components. Similarly, a leaderless community of termites, when large enough, can build a structure that looks planned although no individual termite knows the plan. The structure emerges from group activity. Components or modules need not be physically present simultaneously; they may be presented serially, in rapid succession, and still be effective. Emergent properties appear when a system reaches threshold (an oxymoron for the threshold simply means when the system produces the property). A system is an aggregate of components the output of which is, or is analogous to an emergent property. Threshold can result from the addition of a distinct but critical agent, the accumulation of a critical number of identical modules or as increased activity such as speed or number of responses produced such as the spiking of an assembly of neurons. In the nervous system an emergent property is produced by the activity of one or several sets of neurons (set is used in the broadest sense—a collection of nerve cells and fibers correlated in a given activity) but appears in another domain. This may be a different constellation of neurons, perhaps widely distributed, or it may be nonlocalizable.

Customarily we consider the emergent product to appear on the output side of a system. There is no reason however way such products, if they are tangible, should not appear on the input side and become raw material for later developed emergent properties or products. Indeed, in the nervous system the distinction between input and output is arbitrary. Turnaround occurs at multiple hierarchical levels. In cortex the central sulcus is alleged mark the transfer point between sensory and motor—between input and output. This notion, for the convenience of students, is not sufficiently nuanced. Stimulation of sensory cortex may produce a movement. Histologically typical granular sensory cortex is present in "motor areas". In the human nervous system, emergent properties appear at the highest, most central point of input where the intangible mind formulates the programs for the tangible output of movement and the intangible esthetic structure of performance.

Emergent properties in the nervous system are like a Russian doll-a series of embedded functions linearly connected, each of which produces an emergent property, to culminate in the psychophysical. So ions produce voltage (and current), voltage produces spikes. Threshold for a spike constitutes the simplest neural emergent property. Increased stimulus strength or duration causes increased voltage of the graded generator potential-the analogue of an increased number of modules. A train of spikes if sufficient can produce another and different emergent spike train on the far side of a synapse, a consequence of synaptic currents. Spike trains presumably convey information, and ultimately experience. The conversion of a stimulus to a spike train is often termed encoding, implying that the spike pattern represents the input information in a mathematical isomorphic relation. Attempts to break the code have been unsuccessful emphasizing that the evidence for such a relation is inferential at best. There are several reasons to question a one to one translation between spike train and input information. It would be necessary to show that a train crosses a synapse unchanged or if changed postulate a different post synaptic code. Interestingly, a second identical stimulus to a nerve shortly following a first initiates a different spike train from the previous one. Perhaps this is because the nerve is different as a consequence of the first burst, as can be demonstrated with post tetanic potentiation. Or perhaps as Gertrude Stein would have us believe with "a rose is a rose is a rose" the second stimulus (rose) differs because of the context created by the first stimulus (rose).

Perhaps there is no code. The function of the spike train is to indicate activity and intensity, the former by the presence of the train, the latter by the frequency of the spikes (number per unit time). Intensity could also be increased by parallel processing. Increasing the number of nerves carrying the information at the same time would increase the intensity. What had been considered code would relate to location of termination of the neural activity. Impulses that arrive at thumb area of cortex would stimulate the thumb no matter what the pattern of arrival. The spike train would be an initiator of information; the specifics of the information would reflect the area of arrival.

The transition of spike train to experience is a qualitative transformation. Until then the entire process had been physical with threshold determining the physical changes—the emergences. The last step, the conversion of the physical to the psychophysical represents a qualitative change, a transfer of domains that comprises genuine emergence. The adequate stimulus (Sherrington's term) represents an emergent property in reverse; its initiator. At the simple level a wavelength (for example), a physical entity, is converted by an appropriately tuned receptor into a physical event, a train of impulses. Following an appropriate chain of transmission the stimulus is converted to a new emergent property-a color "seen" by the perceptual mind. For the observer of the world there are no mind-independent properties (as there are in analytic philosophy). There is no objective color, no psychophysical entity, until the final event in the neurological chain—the "awareness" or "consciousness" of color. But there is a stimulus such as the reflectance of a wavelength that initiates the neurological series of events that can be viewed as the source of the emergent property.

The great mystery of the mind is how the conversion of the physical to the psychophysical takes place. How do electrical impulses of nerves become the sensations of mind appreciated by the self? Psychological states are produced by physical events on which they are dependent, but the psychological states—sensations arising in sense organs—are distinct from the physical manifestations that are a coded analogue of the experienced feeling. The feelings are an emergent property, an aspect of mind which itself is an emergent property. This great mystery is no more mysterious than the conversion of hydrogen and oxygen to water (which is, in fact, a great mystery). Emergent properties may be viewed as the most basic function of the central nervous system.

Chapter II

Emergence Of The Psychophysical

If a function of mind is to convert the neurological code to a psychological analogue then, in humans, mind exists from the lowest central nervous level to the highest, for sensory input is felt (in some sense of the term) even at spinal levels. With ascent input is increasingly shaped; unstructured sensations become structured perceptions. Each has its psychological counterpart produced at each level (for example thalamus for sensation, cortex for perception) so mind exists at each level, becoming increasingly expansive as it ascends to culminate in conception. Perception relates to stimuli (and their memory) in the environment. Conception relates only to memory—the mind stimulating the mind.

Mind begins in memory, particularly mind that deals with percepts and concepts, because the actual event is gone just as soon as it is received. It must be retrieved from storage. For our purposes there are two categories of memory: declarative and procedural. Declarative, which can be verbalized, can be further divided into episodic and semantic. Episodic deal with events from life and is largely based on percepts—real or imagined. Semantic deals with abstractions—words, numbers, the meaning of things—and represents concepts. Procedural memory is mainly motor memory for such procedures as walking. Once learned and stored it becomes implicit, automatic and difficult to verbalize.

Declarative memory has its analogues at all levels of the nervous system (without the declarative function of verbalization). Sensory input evokes neural output, the peripheral expression of which is most often motor (in the broader sense). Sensory input at low levels is analogous to declarative episodic memory; it represents something acting on the body. Motor output—movement or resistance to passive manipulation termed tone—is analogous to the conversion of the neural code to its psychological expression at high levels—shape, color, texture and the like. What is stored in procedural memory is the algorithm or program for the required movement. The conversion of a spike train to muscle contraction is the parallel on the motor side of seeing an object or its attribute in the sensory sphere.

Although conversion of the physical to the psychophysical occurs at each level of the nervous system the term psychophysical should perhaps be reserved for the highest level—the one at which verbalization of a phenomenon can occur. A noxious stimulus in a spinal patient induces withdrawal without the feeling of pain. But the withdrawal suggests the delta A and C carrying pain fibers had been converted to something other than a spike train (whether it is called pain or not). The coded noxious stimulus had been decoded. At the thalamic level pain can be felt and its presence acknowledged, but psychophysical conversion of more complex structures or events requires cortex. Cortical lesions may destroy the ability to convert the physical to the psychophysical (or perhaps destroy the physical basis for the conversion) providing inferential conclusions about the site and even the mechanism of conversion. Visual agnosia is an outstanding example. With certain lesions of a high level peristriate visual cortex the patient behaves as if, and may attest that, vision is lost. But stairs can be mounted, doors can be opened, thresholds crossed and if thirsty a glass of water can be reached for and raised to the lips. What is operating visually at one level cannot be "seen" at the next higher level implying that the psychophysical "seeing" occurs in some relation to the injured cortex. The function should not be localized to the site of the lesion (it may be distributed) but that location is a participant in the distributed function of conversion of the spike train to whatever it represents.

Chapter III

Organization Of The Nervous System

The nervous system is a true system composed of modules that are organized in conceptual layers. The fundamental unit is the grey nerve cell body with its white appendages called axons and dendrites. The axons transmit, the dendrites receive and the coded language is the electrical impulse called the spike. The junctions where cells connect are called synapses. Here information from one cell—the presynaptic member—is transmitted to the next cell in line—the postsynaptic cell. Collections of pre and post synaptic cells form assemblies, a dynamic group able to change organization rapidly, so a given cell or cell group may participate in more than one assembly. Modules are composed of cell assemblies and receive information from the environments. At each ascending level that reflects increasingly later evolutionary acquisitions the information is processed and shaped. It becomes increasingly refined and its output action increasingly individuated. The mass reflex of the spinal level (a triple flexion response) is converted to movement of a single limb at brain stem and to a delicate movement of a single digit in cortex. Centripetal information is variously labeled afferent, sensory, perceptual and conceptual as it climbs the neuraxis. Afferent operates at the lowest level without coming to conscious awareness until impaired. Vestibular and position senses that function at brain stem and spinal levels are examples. Sensory is a term reserved for unstructured sensation. Appreciated at a thalamic level pain is a good example. Pain, an early evolutionary protective signal also operates at a spinal level but without awareness of the input, without sensation, without "pain". The next input level is cortical, where thalamic sensation is constructed into percepts. They remain coded until converted to psychophysical form by the perceptual mind. Shape, color, texture and all other characteristics of the environment appear as signs—mainly icons. Percepts have meaning. The meaning may be literal, may be metaphorical, may be contextual (either internal or external) and may be accompanied by emotional overtones. The abstraction of the percept, in all its meanings, is the basis of conception which is the penultimate neurological level to be transformed by conceptual mind into arbitrarily chosen manipulable symbol systems that (unlike signs) have no aspects of the referent. They constitute and are a code; simply a different code than the original input.

Dedicated cortical areas are a late developmental achievement. Early developed cortex in the individual is assigned only general function and is extremely plastic. The anatomic module is physiologically flexible. Visual cortex in the early life blind (consider pianists George Shearing, Art Tatum and Alex Templeton) becomes responsive to sound. The psychophysical is not a compulsory product of the physical anatomy, as shown by the recovery of language following a left hemisphere lesion of early childhood. Language is now "transferred" to right hemisphere. This plasticity, most apparent in early life, is not limited to cortex as shown by

Bechterev's nineteenth century ablations followed by recovery of vestibular activity in dogs.

Cortical development is partly epigenetic and occurs predominantly in critical periods as shown by amblyopia ex anopsia. The requirement for input may extend as far peripheral as the receptor. The great variation in infantile environmental exposure fosters great variation in neural development and may relate to such entities as developmental dyslexia, congenital prosopagnosia and so called tone deafness or the inability to carry a tune. These types of developmental variations result in limitations of percepts that in turn limit concept formation and account for the great variation of individual minds.

On the efferent side all overt output is converted to motor activity. Motor activity, whatever its nature, manipulates at least one of the three environments. This is as true for glandular secretion as it is for performance of a Bach partita. What is not overt nor known to others is the internal performance of perceptual and conceptual minds. This is as much a motor performance as any muscular event. Perceptual sensation (psychophysical) is drawn from episodic autobiographical memory as conceptual abstraction is drawn from semantic memory. These two declarative memories are accompanied by procedural memory, the memory of motor performance. This may be for an automatic movement such as walking or for a never previously performed gesture made up of fragments of familiar bearing. (Note the similarity to language; the synthetic ability of syntax). The movement is drawn from procedural memory; a new syntax is created there for a new movement. This is projected onto one of the three environments as performance. This is the analogue on the procedural output movement of what in perception remains internal (the translation of the perceptual neural code to the psychological sensations) and in conception may or may not remain internal (the conversion of internal language to external speech by way of procedural memory). All high level internal states, whether or not they have achieved psychological expression must pass through procedural memory on their way to the periphery; lower levels can initiate movement without contacting procedural memory but the so called reflex movements are coarse and serve mainly protective function. As mirror neuron discharge indicates, perception of an action—movement—induces activity without motor accompaniment. Even an incomplete percept (monkey) such as reaching behind a screen by the examiner for a hidden reward induces mirror neuron activity. Surely this is a demonstration of conceptual mind and indicates its presence in monkey. Performance in absentia is the output expression of perceptual and conceptual minds.

The unfamiliar paradox in all of this is that there is no separate observer looking at this output of brain. The brain is observing itself. The brain, part of the internal environment is being observed by the brain considered distinct from the environment. Furthermore, this structure of internal environment is largely concerned with the external environment. Where is the self in all of this? To suggest that self is in brain would be acceptable to most. To suggest it appears at the highest level would seem reasonable. To propose it is widely distributed, present at all levels, would seem discordant. Once again it is a matter of how things are measured. If self is (Webster) "The ego; that which knows, remembers, desires, suffers" that sounds like a description of high level mind. If it includes low level critters that withdraw from noxious stimuli (we don't know if they "feel" pain) that is an argument for distributed function.

The current concepts of top-down or bottom-up organization are often understood to be linear. They are better conceived as circular, or better still as ascending or descending spirals. A great deal of interaction occurs at a horizontal level often with input from above or below. There is feedforward and feedback understood to mean derived from another link in the chain. In most non-neurological systems feedback, which can be positive or negative is generated within the module. In a mechanical rotator with hinged wings (for example) the faster the rotation the higher the level of the wings; with the higher level the slower the rotation—negative feedback. In biological systems, depending on how the module is defined, the generation of feedback may occur within the module or external to it. If the single neuron is understood to be the module most often feedback comes from without. It could be argued this is not truly feedback; it is no different from information transmission anywhere in the nervous system. But output from the first cell to cause the next cell in the chain to fire back and amplify or decrease the firing of the original cell gives it the character of feed control. The Renshaw inhibitory cell that projects onto the alpha motor neuron gives an example of feedback arising external to the cell. It receives an excitatory collateral from the alpha motor neuron and projects an inhibitory fiber back on to it. If the cell assembly is conceived as the module feedback may arise within. Recurrent collaterals exemplify structures concerned with indigenous feed.

Feedback and feedforward are but one of the mechanisms to induce modification of neural activity at each level ultimately to shape the final product. New information is introduced from above and below. This principle of organization is seen even at the lowest level. As an example take muscle, productive of all movement (behavior), the only overt manifestation of nervous activity. It is activated by the spinal alpha motor neuron. This receives segmental input from the I A dorsal root afferent reflecting the state (length) of the muscle spindle, controlled by the segmental gamma motor neurons that receive afferents from cortical and subcortical regions.

Input and output at concrete levels deal mostly with the contralateral field. With greater abstraction homolateral fields are included. Vision from the same side, as well as crossed is represented in temporal lobe. Language does not stop at the midline; we can read in both visual fields. More and more unification occurs at higher and higher levels until ultimately we have one mind and one self.

Chapter IV

Evolution

In evolution the nervous system developed from the bottom up—spinal cord before brain stem and cerebellum, then on up to cerebral hemispheres and cortex. It is not as if each level were completely shaped before being succeeded by the next. Modulation and development continue even as the next higher levels are applied. The early reticular formation of spinal and brain stem is supplemented by lemniscal ascending and descending tracts all the way up to cortex. Behavior, in evolution, occurs independently at each level, to be modified by descending tracts as higher levels develop. What was "reflex" behavior at a spinal level becomes "voluntary" behavior by the time (in evolution) cortex is acquired (requiring definitions of reflex and volition). Behavior means movement that serves an immediate purpose such as acquisition or translation, but also communication (more than just speech or body language). As Sherrington noted mind is always an inference from behavior. What may also be true is that mind is an evolutionary outgrowth of behavior and ultimately of movement.

The reflex is the behavior building block out of which all higher behavior is built, just as the spinal cord is the neural structure on which in evolution the remainder of the nervous system is erected. (In this and all other discussions, the terms higher and lower are used in an evolutionary sense meaning later and earlier evolutionary acquisition respectively). At the spinal level the reflex is a monosynaptic arc which supplies the basic movements of approach or withdrawal. The arc consists of an afferent fiber that conveys the stimulus and an efferent fiber that initiates the muscle movement of response. The reflex response is stimulus bound; it must occur. As the reflex is modified by the addition of input from higher levels it becomes increasingly facultative until it can be entirely suppressed. With the acquisition of higher levels the stimulus can become less concrete until, at the highest level, it becomes a product of mind, a "voluntary" event, a manifestation of will. The term tropism has been introduced to be distinguished from reflex perhaps with the distinction that tropism implies growth that may produce movement and reflex implies movement without growth.

If the term stimulus is defined (as it should be) as an input signal that evokes a response, then clearly the response defines the stimulus; that is, no response means the environmental physical event did not serve as a stimulus (allowing for response delay or facultative deferment). So the term subliminal stimulus becomes an oxymoron and points up the problem of how to measure behavior and of operational definitions. As will become evident this is important in a discussion of mind for stimulated behavior may be "mindless" and therefore considered "subliminal". So called subliminal stimuli have been long known to evoke behavior. All this means is that the presence of the stimulus cannot be verbalized. The stimulus—because it is too brief or too weak—does not reach the language—mind complex, but it does reach lower levels and can evoke non verbal behavior at these levels. The stimulus remains a stimulus because it produces a response. The question becomes how a response is measured. If behavior becomes a criterion of response and if mind is an inference from behavior the conclusion that behavior is a criterion of mind is not justified. Unless mind be redefined in the recursive sense to exist at every level of the nervous system gaining increased complexity at each higher level. In this sense non-human animals have minds, as their behavior demonstrates.

Evolution left its mark on the human nervous system. From the early slow transmitting reticulate organization it added rapid transmitting lemniscal pathways for sensory transmission to thalamus and thence to cortex and for rapid motor excursion by the corticospinal tract from cerebral hemisphere to spinal cord. Before cortex developed in evolutionary time lower structures were being elaborated. Brain stem with cranial nerves and cerebellum appeared early in history to be succeeded by subcortical and cortical development. The lissencephalic cortex expanded to the point it turned in on itself by producing gyri and sulci to allow the surface area of the mantle to expand many fold. Although we depict this as a linear serial occurrence, lower levels were expanding in parallel with the development of higher levels. The anatomical attainments were accompanied by physiological growth and the advent of new functions. The cortex continues to evolve with function developing in part epigenetically in response to use. As information arrives at cortex from spinal cord (pain, touch, temperature, position), thalamus (vision) or directly (smell) it recapitulates in milliseconds what evolution developed in millennia. The course of the nerve impulse is the course of evolution. Note the privileged direct position of olfaction, the oldest of the special senses. It goes directly to rhinencephalon, the oldest cortex.

Like spinal cord it wears its white matter of myelinated nerve fibers on the outside. Later developed cortical areas have grey matter of nerve cell bodies external. Olfactory cortex has fewer nerve cell layers than later acquired cortical areas. The olfactory nerves are a direct outgrowth of brain (like optic nerves) unlike other nerves that arise peripherally and connect to brain. The evolutionary recapitulation is what accounts for various levels of neurologic function even for events that reach the highest mental levels where they are shaped and refined by feedback from the same or other levels. With ascension the program for an event becomes increasingly abstract to culminate in perception with its attributes and qualities translated to psychophysical sensations. On the efferent limb programs become increasingly concrete with intention (contingent negative variation) preceding ultimate muscle contraction.

The time it takes for a sensory stimulus in the environment to reach cerebral cortex can be calculated by the technique of evoked potential for visual, auditory and somatosensory stimuli. Time to subcortical waystations can also be observed for auditory stimuli. Electroencephalographic methods demonstrate the cortical appearance of the wave form that indicates arrival. Waves are described as negative (N) or positive (P) and the time in milliseconds. It takes about 20 milliseconds for a stimulus to reach thalamus; later waves reflect cortical activity. They may be as late as 300 mS preceded by earlier cortical response. This suggests several levels of cortical participation with serial hierarchical relation. P300 has been correlated with decision making so may be an electroencephalographic manifestation of mind. This is emphasized by studies on the readiness potential of contingent negative variation. Before an individual "decides" to perform a movement (a manifestation of mind or will) an electrical potential appears about 300mS (more or less) to be translated. The translation is what the self is aware of and is spoken of as consciousness or will while the earlier electrical phase—presumably generated in a separate cortical area—is defined as unconscious. This progression simply reflects the evolutionary itinerary of the nerve impulse. The paradox is that in these experimental procedures the readiness potential must originate in mind. The subject is instructed in advance to move a finger at will or to perform an instrumental act on signal. Mind is projected downstream to initiate the contingent negative variation. Whether this occurs in non experimental settings is not known. It would not be surprising if it did. It would simply indicate that in "mindless" non human animals neurological preparation for performance exists; what is observed in humans is an evolutionary residual (like many other neurological functions).

How far downstream mind is projected to unleash lower levels is conjectural. We know that in structured experiments the cortical pyramidal cells (output of which is carried downstream by the corticospinal tract) fire in advance of movement and cease firing when spinal motor neurons initiate muscle contraction. Here is a case where "willed" movement is projected all the way to the periphery in practically a single leap to the surrounding environment. Can the same be true on the sensory side? Does the mind construct and color the environment or merely build and paint the model? Is there a philosophical difference?

Chapter V

Environment

A stimulus is the biological effect of an event in the environment. The environment consists of three concentric spheres: the external world, the intermediate body wall and the internal innards of viscera. These three environments are served by three sets of afferent neurons: exteroceptors, proprioceptors and enteroceptors respectively. In the embryonic nervous system are three concentric layers of cells; the germinal, mantle and marginal layers from within out. These become, in the mature nervous system the central autonomic substrate (hypothalamus and intermediolateral horn cells), the subcortical nuclear layer (thalamus and basal ganglia) and the cerebral bark (cortex). These three derivatives deal respectively with the functions of the viscera of the interior environment, the movement and posture of the intermediate body wall and the outside world and reaction to it. Information from each layer of the environment and from each layer of derivatives is conveyed to cortex to arrive ultimately at multimodal then supermodal areas to be scrutinized by mind.

Much of the information about the outside world is received at a distance. It is transmitted by waves and quanta—photons—and impinges on distance receptors—teloreceptors. Teloreceptors are clustered in dedicated regions—the retina, cochlea, olfactory epithelium, taste buds—that serve special senses. Teloreceptors are tuned not only to a specific modality but to certain frequencies of that modality (vision and audition). This may result from feedback to the initial site of input as with the stapedius muscle for sound or from selection of the next member of the neural chain such as the cochlear hair cell for pitch. At the peripheral stage control and shaping of input often begins.

Receptors for common sensation—pain, temperature, touch, position—are more widely distributed over the entire body. Pain is rather different from other sensations. In evolution it is present early and distributed widely in the nervous system. Two fiber systems convey it. One type of pain—probably historically the earlier—is termed protopathic. It is poorly localized, diffuse with a frightening affective quality that is typified by visceral pain—intestinal gripes, menstrual cramps. The second type called epicritic is bright, well localized and less frightening. Pain is transmitted centrally by the slow C fibers for protopathic and the faster delta A for epicritic. (A and C refer to fiber size; A is the larger). There are no specialized receptors for pain as there are for other common sensations; bare nerve endings receive incoming painful stimuli that are brought by delta A and C fiber impulses to the central nervous system where they are distributed widely to reticular formation and focally to two sites in thalamus by a lemniscal system—a later evolved set of tracts than the exclusive reticular organization of more primitive organisms. Pain operates at many levels to produce and shape behavior but, as with other modalities that may also operate at low levels, it is not felt as pain until it reaches a later developed thalamic or cortical analyser.

For convenience input to the nervous system may be organized in a conceptual hierarchy. At the first level is afferent: impulses that ascend and are effective in controlling behavior without awareness by the organism. Vestibular and proprioceptive input (major controls of posture and movement) are examples. Afferent impulses that reach awareness but remain unstructured may be termed sensory. Pain, touch, light, noise are examples. When multiple sensations become structured percepts are formed. A particular combination of shape, color, texture and size forms a percept (of perhaps a table). Although the neurologic coding for aspects of the percept (color, shape, texture) are present in the percept, uncoding (yellow, square, smooth) requires what might be called the perceptual mind. Abstraction of many percepts of an object (many tables) creates a concept. The conceptual mind classifies and manipulates the abstractions of the concepts. Multi-modal areas of cortex, late evolutionary acquisitions, collect multiple sensory input, probably in parallel, to create percepts. Later evolutionary development of cortex becomes the substrate of mind. Neural codes from percepts formed in multimodal cortex and their abstractions serve as input. Conversion to psychological analogues occur as output in frontal and temporal cortices connected one to the other by the uncinate fasciculus which is particularly robust on imaging studies in individuals with superior autobiographical memories. These highest cortical areas, that include a location of importance in the production and comprehension of language, can be conceived as the neurological substrate of mind.

Chapter VI

Percepts And Concepts

The standard teaching about perception is that there is a throughput of individual sensory modalitites to perceptual cortex where they are combined and assembled into a percept. There are multiple problems with this approach including the binding problem. How are the various aspects of a percept put together to form a whole? Presumably this is done at the neural rather than at the psychophysical stage but there is no compelling evidence either way. The nervous system operates on sensory input at all levels, from the most peripheral to the most central. At each level there is feedback and feedforward. From the multiple array of stimuli to which the peripheral receptor is exposed it responds only to that to which it is receptive or tuned. The retina, for example, is responsive only to light. But not to all light. Individual cones respond to either red or green; rods respond to blue. Response is to wavelength, not to color and as is true of all receptors, is graded. This generator potential is transmitted to bipolar and retinal ganglion cells that produce spikes. Retinal ganglion cells respond to stimuli of biological significance (feature detectors such as frog bug receptors)—a principle at many levels of neurological activity. Spots of illumination are organized into bars of orientation at an early cortical level; individual simple cells respond only to a bar of light with a particular orientation. At each ascending cortical level more components are added to arrive at multimodal cortical fields with bilateral representation. From each level there is feedback to lower levels to help shape input. There is a larger projection from cortex to lateral geniculate body (thalamus) than from lateral geniculate body to cortex. By the time of arrival at the multimodal regions parallel processing has the coded percept practically formed. Its pieces need to be assembled, like a jig saw puzzle (or like television pixels) but like a jig saw puzzle each piece carries multimodal information that contributes to the completed picture. The binding problem consists merely of fitting the component pieces together. And like a jig saw puzzle when a sufficient number of pieces has been assembled the unfinished picture can be completed conceptually; it can be inferred to fill in gaps in perception.

Percepts often have a emotional quality that is not an intrinsic component of the percept. That is, it is not transmitted by the sensory channels that comprise the pathways of sensory input of which the percept is constructed. The emotional accompaniment is not an obligatory component of the percept. The sensory structure of the percept cannot be changed without altering the percept; the emotional contribution can because it is not an intrinsic part of the percept. It may reflect something unrelated—an anticipated event or a previous experience called up by the percept. It is a component of the percept that operates, and is transmitted, in parallel with the sensory structure of the percept.

Perception, which represents the periphery, is stimulus bound. It requires input from the environments to develop. Concepts, abstractions of percepts are stimulus free and can occur independent of the tangible environmental stimulus.

Concepts are also language independent although they require a symbol system like language (or mathematics) to be expressed. One can conceive of a lemon (for example) without words—its shape, its color, its texture, its aroma, its taste. This is done mainly by revisualization. Vision is the major sense contributing to percept formation. Other senses contribute but are not as manipulable. Sound—music—may be reheard in the mind (and may be reactuated) but smell, taste and common sensibility are confined to the presence of stimuli. Revisualization makes the transition from perception to conception by way of an icon. The transition to a symbol system is by means of a sign.

In an individual human there are two separate minds dealing with perception and conception. One, the earlier evolutionary development is the perceptual mind. It is presumably shared in varying degree by all mammals. It is concrete, requires ongoing persistent stimulation and consists of signs. There are mostly iconic (humans are largely visual creatures) but other types of signs are included. It is created by, represents and is controlled by the outside world. The two other environments contribute little to perception. We are generally unaware of our viscera in the internal environment, and input from the middle environment of the body wall is usually unrecognized until something goes wrong with position or vestibular sense. Pain—a primitive sensation—is the one exception that can make the two environments apparent. But pain is an unstructured sensation, not a percept. The external environment cannot be manipulated mentally. It can be physically changed but mentally must be perceived as it is. The perceptual mind applies the psychological aspects to the coded neurological input. Attributes are realized and can be felt, seen, heard, smelled and tasted.

The conceptual mind is a higher order later evolutionary development that is presumably uniquely human. It is abstract, conceptual and manipulates symbols. It is independent of external stimuli so can create environments that never existed. It is the basis of imagination (that can be done without images). If revisualization is used it's a sign carried over from perceptual mind. Language is the major symbol system generated by the conceptual mind and may function as a source of conceptual mind—an outgrowth that serves as a source of input. The conceptual mind is the most private part of the nervous system. The perceptual world, and therefore the perceptual mind is shared with others. The conceptual world, created by the conceptual mind, is unique. It is the composite of abstractions of the life of the individual, different from the lives of every other individual so no two conceptual minds are alike.

Chapter VII

Language

Language is a major component of mind. It is an internal symbol system to be distinguished from speech which is its external expression. It is created by mind and paradoxically helps create mind. It is not the only mental symbol system; numbers, mathematics, music all exist but language may be the broadest and most easily manipulable. A symbol is an arbitrary indication of that which it represents, is not iconic nor produced by the referent (as a footprint would be a sign but not a symbol of a passing animal). Language, like all conceptual systems, is stimulus independent, unlike perception which requires the presence of the stimulus. This gives language great flexibility. It can represent objects and events that do not exist and never have existed. It represents the world but can create imaginary worlds. Like all components of mind it models reality. Unlike perception which denotes isomorphically the conceptual model (whether linguistic or not), an abstraction, is isomorphic only insofar as a connotation is isomorphic.

Language has many functions other than modeling reality. Communication is usually offered as the major function with the semantic aspect of the words conceived as the vehicle. But the prosody, the tempo, the pauses, the associated gestures, postures and facial expressions may be more informative than the words themselves. What is not said may be more revealing than the language used. Miscommunication may be an intentional function to transmit disinformation. We tell lies all the time; how hard it would be to do this without language. Then there is phatic communication. It says we are there, we are participating, we are listening. It does this with words but generally independent of their meaning. They just serve to make listening noises. Language is used to manipulate and control. How often does the word "don't" appear, particularly in our dealings with the young. We talk to ourselves—perhaps for security, perhaps for companionship, perhaps for motor programming. A great deal of performance is conceived prior to being executed. Some of that conception is in words. A well dressed man enters a crowded elevator in a downtown office building and says to no one in particular "eleven" just before he pushes the button. Finally language plays a role in thinking. Not in all thinking, for perceptual thinking is largely iconic and non verbal. But formal organized thought is often linguistic with subvocal words.

The close association between language and mind is demonstrated by a study of eye movements in patients with aphasia. Using a nonverbal paradigm (no instructions are given) eye movements exploring a set of pictures were compared in aphasics and controls. Results indicated that aphasics did not comprehend what there were looking at. Whether the defect in comprehension was the result of the disorder of language or whether the abnormality of language reflected the abnormality of mind (or whether both indicated a third common cause) cannot be decided on the basis of the study but the important point is that aphasia is more than just a disorder of external language; it implies a disorder of internal language with an associated disorder of the model of the environment that constitutes an important component of mind. In the nonaphasic the connection between language and mind is intimate: mind is expressed by language and language expresses mind.

Chapter VIII

A Model Of Reality

One of the major activities of the nervous system is to model reality. Multiple inputs received by transducers are converted to transmitted signals, combined, analyzed and viewed. The neurologic model is not isomorphic (in the biologic sense; in the mathematical sense, of course, it is) with reality and may not be a faithful representation but is the only way reality can be known. Without it there would be no external world. Much of the external world and its properties are created by the model which is built neurologically level by level—the representation, rerepresentation and rererepresentation of Hughlings Jackson. This is an active process in which the nervous system operates on the input to shape and structure it, to create the model. The neurological activity is bidirectional; not only is there ascending input but there is descending shaping of that input to determine the final product. The input that stimulates a given receptor is quite different from the effect of the input, and demonstrates the requirement that there be a nervous system to analyse and decode the input. What is decoded may not be part of the code but be an analogue. Take color as an example. In the environment an object has no color; it merely reflects light of a certain frequency. That frequency is responded to—stimulates—certain retinal receptors that generate a potential passed along to ganglion cells, lateral geniculate nucleus of thalamus, occipital cortex and forward. It is combined with other visual input and ultimately analyzed in a supramodal area of cortex and described as "red". That "red" is a product of the nervous system (the conversion to the psychophysical) and did not exist in the environment. It was not present in the object, in the light frequency, in the retinal cones, ganglion cells, thalamus or occipital cortex. There was no "red" in the spike trains that transmitted impulses from one level to the next. There was no "red" until the highest level of the nervous system "saw" it and knew it saw it. The coded red color could have functioned—have had an effect—before that but it would not have been recognized as a color. This is true for all qualities ascribed to the external world—texture, temperature, flavor, taste. They are qualia supplied by the nervous system that do not inhere to the objects of the world. So it is with the entire outside world. We do not know how it is; we only know how we model it.

How we model it is more a function of the nervous system than it is a statement of reality. The reflection of reality is a result of how the nervous system was put together in its evolutionary development. Because the human nervous system is a continuation of evolutionary development in other animals we can all operate in the same environment.

In evolutionary history supramodal assemblies are a late achievement. They are located in frontal and temporal regions, receive coded input representing percepts and concepts and decode it. Output consists of attributes and feelings that these assemblies can "read" to become the substrate of mind for percepts. This is done with iconic signs to which, from emotional memory (wherever that is) feelings are added.

Attributes and feelings do not exist in the same sense for concepts in which instance the readout is converted to abstract symbol systems, such as language, that can be manipulated and transmitted unlike the signs of the perceptual readout that must first be converted to concept and symbol to be communicated.

What makes this formulation of the psychophysical manifestations of perception counter-intuitive is the expression of the perceptual components (decoding) and the psychological recognition of it are a unitary phenomenon—done by the same "person". But compare it with language or with any other motor behavior. Language is the expression of conceptual symbols by the same mechanism that encodes them. If you believe the classical distinction between Wernicke's and Broca's areas that is simply adding another set of neurons to the assembly for expression. But Wernicke's area alone is the analogue in the conceptual sphere of the perceptual psychophysical mechanism. Any motor phenomenon initiated in cerebrum drawing on procedural memory is a similar phenomenon: encoding, decoding and performing by the same sets of neurons, the same brain, the same person. Its output, the realized activity is an emergent property displayed by the nervous system in the external environment.

What occurs on the motor side originates in procedural memory where a program is extracted for a specific (dance the tango), a general (play the piano) or an automatic (walk to the post office) function. The program is encoded and transmitted across multiple levels until the final common pathway, the spinal alpha motor neuron, from which it is distributed to muscle where the code is transformed into action. Small motor units innervated by small neurons of lowest threshold are recruited first with larger ones following if greater strength is needed. The conversion of neural code to "action" is the counterpart on the output side of "feeling" and "attributes" on the input.

We are accustomed to think the entire interaction with the environment on the output side of the nervous system is motor. But the model of the environment created by the mind, its psychophysical form, is projected on the environment by the efferent nervous system as well. How closely that psychological model accords with the physical environment—how what we see is the way the world looks—is a philosophical problem that probably will never be solved. But the relation between the mind's model and the real (physical) world can be explored in two pathological states. In one the environment is absent but the model is complete. In the second the environment is complete and the model is lacking. In each some modalities—vision particularly—disaffirm the sensations and the convictions but the mental persuasion persists.

In the phantom limb syndrome an absent member—a finger, a hand, an arm or a leg, a breast—may be felt present although seen and known to be absent. The phantom may be painful or not. It may be felt to move. It may adopt impossible postures. It is suggested to result from input from the residual nerve much the way striking the "funny bone" projects a sense of tingling from the fingers. The disparity between actuality and the psychological model, the product of mind, is insufficient to alter the model.

In the second situation the environment is complete and the mental construct is deficient. In patients with a lesion of right parietal lobe (parietal lobe is allegedly important in percept formation) the patient's left arm is recognized visually as an arm but not as the patient's by the patient. Often is assigned to the doctor who may then show his two hands. Still the patient ascribes the denied left hand to the doctor. An impossible conversation follows: the patient agrees that most people have two hands, the doctor must have two hands, the patient's left hand therefore cannot belong to the doctor. Then the process starts over. Shown the left hand the patient denies ownership and says it must be the doctor's. The environment is complete, the model is deficient.

These two pathological conditions suggest that in the normal the psychophysical model of the environment is not a passive internal state but like the motor system it too is an efferent system projected forward to the surround to manipulate the environment as do the motor system and external language. It is an active system that shapes the environment on the output side just as the environment is shaped on the input side.

Chapter IX

Mind

Mind: an elusive entity. Hard to define. Harder still to understand. Webster is of no help. "The totality of conscious and unconscious processes and activities of the organism". This only raises the further problem of consciousness. For many years mind was considered independent of body. Descartes, whose cogito ergo sum created centuries of subsequent mischief could be understood to mean "my mind is my self". Taken in this sense it follows that an understanding of mind will provide greater understanding of the concept of self. Now it is generally agreed that mind is a product of nerve activity so perhaps one way to approach the problem is to study the nervous system. We think of mind as essentially human, but non human animals, particularly mammals, give evidence of mind. Dogs wag their tails when happy, cats purr when content. Happiness and contentedness are states of mind. Perhaps we are anthropomorphizing when we assign happiness and contentedness to behavior. But cows locate and recognize (a function of mind) the barn at night and geese find their way home.

There seem to be two kinds of minds. Both require cortex to function. Non human animals have one, humans are endowed with both. The earlier evolutionary endowment is the perceptual mind drawn from perceptual memory. It creates a model of reality that may not be isomorphic with the exterior world. Percepts are constructed of sensory input (this does not mean sensation) and have attributes and qualities. Percepts are things—objects and events (nouns)—while attributes and qualities are characteristics (adjectives). Percepts are stimulus dependent; they require the presence of a stimulus. Much of stimulus input for objects (nouns) and action (verbs) is visual so it is of interest that there are two visual input pathways: a ventral path for objects and a dorsal one for movement. This perceptual mind also exists in lower (in the evolutionary sense) less evolved animals.

The second kind of mind is exclusively human so far as we can tell. It is a late evolutionary occurance, is stimulus independent, consists of abstractions and meanings, is intimately associated with language and thinking and can create objects and actions that do not actually exist (as may occur in dreams) in its model of reality. It allows the use of symbol systems and accounts for creativity. It is the conceptual mind.

Both minds are based in declarative memory—perceptual and conceptual memory. Declarative memory is composed of two components: episodic or autobiographical memory and semantic memory. Autobiographical memory deals with objects and events that have been perceived. It could be termed perceptual memory. Semantic memory is concerned with abstractions and meanings; it forms conceptual memory.

All these functions are coded in neurologic activity. They are physical. The qualities and attributes of percepts (and the percepts themselves) and the abstract manifestations of the conceptual system are psychological. A major unresolved question is how the conversion from physical to psychophysical occurs. Psychophysical means sensations evoked by physical stimuli; the difference between luminance (measured with a photometer) and brightness is an example. The neurological representation is devoid of sensation. How and where is it translated? The proposal is that there are regions of cortex in frontal and temporal lobes, later acquired in evolution than perceptual and conceptual areas that translate the physical to the psychophysical. This cortex constitutes the neurologic substrate of mind. It is the critical addition that provides the physical base for the emergent property called mind.

Mind is a readout of all that has neurologically preceded, a summary of conception. It is a late evolutionary development so must appear in late evolutionary acquired neurological structures. Schematically in evolution brain grows forward, up, around, backward and down to terminate in temporal lobes. The place to look for mind then in so far as it can be located anywhere (like many neurological functions it is better considered a distributed function in distributed structures) would be in frontotemporal lobes. What makes this particularly appealing is this is the region on which language is dependent and language is of prime importance in and a strong representative of mind.

The readout of mind is done by the neural structures that create it. This apparent paradox reflects our customary separation of observer and observed. We need a separate concept or agent to analyze the mind so we create "self". If they are unitary then together they have evolved, exist recursively at multiple neurological levels, become increasingly developed as they ascend the neuraxis and increasingly independent of environmental stimuli. The concept of "self", developed at the highest level, is necessary for the awareness of self. Most animals it is agreed, have minds (although how far down the animal scale this goes may not be agreed). It would also be agreed that animals, no matter how advanced have no concept of self although they have knowledge of existence. Studies in monkeys with mirrors are ambiguous and fail to prove a concept of self. But animals know where they are and that they feel hunger and pain—manifestations of a perceptual mind but not of self.

Mind begins in memory. This may seem evident for conceptual or semantic memory, less so for perceptual memory. Perception is a series of instants, and as Hippocrates taught us the instant is fleeting. Perception is like a movie film—a series of static frames. Each needs to be clothed in the attribute of mind; each needs to receive a psychological cloak. Memory serves as the neurological input to a higher level where the output is psychophysical and gives meaning to the perceived world. Memory exists without mind, without the highest level to which to project. It exists in some form at each level of the nervous system, functions there, is projected recursively to higher levels (where it is modified) and underlies behavior at each level. This organization exists in all animals and can achieve the highest and most abstract level called the human mind—a manifestation of more than perceptual input. Semantic memory is an instantiation of the abstractions from the perceptual level—the "meaning" of the percept—now encoded in memory.

According to this formulation memories are stored in assemblies of neurons that subserve the original percept or its abstraction. When the memory is to be initiated the assembly (which is a transient structure, some of its components participating in other assemblies) is reassembled and the percept or concept recovered. The trigger for retrieval resides in the hippocampus but the retrieved assemblies are at the site of original production.

The last evolutionary cortical development provides the critical modules for the emergence of conceptual mind and semantic memory. Perceptual mind—and presumably episodic memory—based on a system of signs is an earlier historical development that allows transition to psychophysical states but is a literal, environmentally bound mental state. Real mind, expansive, creative, free wheeling is uniquely human and emerged with the cortical areas on which language and other manifestations of mind depend.

Chapter X

Conclusion

In evolution behavior came first; mind was a later development and grew out of behavior. Behavior, no matter how complex or refined is always on a motor base. Aspects of the base may not be motor—the esthetic character of a performance for example—but are always an accompaniment. Or a consequence of the motor activity. At the outset in evolution behavior was probably a motor reflex response to a perceived stimulus. How the stimulus was perceived cannot be known. But the stimulus had meaning. Even in the absence of a semantic system the earliest evolutionary stimuli had meaning perhaps determined in part by the internal state of the organism. To a hungry frog a bug is lunch. Early responses were probably ballistic. The flick of a tongue to catch a bug was unleashed with no further control. Once released it continued unchanged until completed. Movement came in preprogrammed packages without feedback control. Even in the absence of mind the program was complicated. A flicked tongue was somehow instructed as to the direction and speed of the bug's flight and the expected time of arrival at the projected site of tongue protrusion. In this formulation

The memories are indivisible because they are a distillate of all that has preceded. This distillate or essence along with all other concepts is projected toward the periphery, perhaps through procedural memory to be activated as motor performance, perhaps to perceptual memory to be revisualized (a motor performance) or perhaps to the next higher level to be "looked at" (a motor activity) on the output side.

All of this is done symbolically. Language is the major contributor and the major means for externalizing the private internal conceptual world. In early individual development language is only used to manipulate the perceptual environment. At about age 7 semantic and therefore symbolic values are developed, implying the development of conceptual memory and mind. The symbols of language are meaningless until endowed societally accepted meaning. Recursion, a principle of neurologic function throughout the nervous system, is perhaps most apparent in language and mind. As the 7 year old mind grows so does language. As the 7 year old's language grows so does mind. These may not be in a linear relation but the interaction of language and mind throughout life signals an intimate and probably casual relation.

The location of mind and self is a quixotic pursuit. The two are conjoined, indistinguishable and probably identical. They are widely distributed in various realizations throughout the nervous system. In so far as any high level function can be localized this is done most easily by lesions. Those lesions that interfere with language (both internal and expressed) only indicate a cardinal area important in the language process. Even if not the site of language localization this is important information both with respect to language and with respect to mind. Late developed frontotemporal regions are the presumptive candidates for human conceptual mind in the sense that they are critical areas either in number or in function to be added to all the modules of mind that have preceded to produce the emergent property of the human conceptual mind.

Ironically the close relation between language and mind was underscored by the slogan of a charitable foundation to raise funds for education of needy college students.

A mind is a terrible thing to waste.

the first manifestation of mind was the development of the corollary discharge—perhaps the first expression of feedback. A projected template of the movement in progress was used against which to measure, by way of sensory information from the moving member, how accurate the movement was in comparison with the programmed plan. Unlike the ballistic movement, adjustment of response could be added to the output. This comparison could be construed as the first instance of behavior looking at behavior and the lowest level of mind—an argument for mind as a distributed function (perhaps dependent on the supplementary motor area from which errors are projected forward to rostral medial prefrontal cortex). This occurred without awareness by the organism, a situation that persisted at many levels, including cortical, as evidenced by so called subliminal stimuli capable of inducing very complicated bursts of behavior by perceiving without perceiving. Mindless behavior is not without mind; it is with mind that has not come to awareness if awareness is understood to mean the ability to verbalize.

Mind, physically created, is actually the psychophysical expression of the physical basis. But "psychophysical" is too restrictive for it deals with sensation produced by physical stimuli, and mind is considerably more than sensations. We are told (Lakoff and Johnson) that "the mind is what thinks, perceives, believes, reasons, imagines and wills". Of these only perception has to do with sensation. Psychophysical perception is the foundation on which the higher activities—thinking, believing, reasoning, imagining, and willing—are erected. Perception comes earlier in evolutionary time, earlier in individual cortical development and earlier in receipt of incoming neurological information than the higher later developed conceptual cortical areas. What the perceptual mind injects into perception is the contribution of the senses—particularly the special senses and most particularly vision: shape, color, brightness but also sound, fragrance, taste and feeling. This is done iconically and without ability to communicate until it is transferred to conceptual regions where the percepts can be assigned symbols that can be shared. The perceptual world is an internal private world but the percepts of the real world are shared by other human and nonhuman animals, particularly mammals. It is a comparatively early evolutionary achievement. In contrast, the conceptual world, the highest evolutionary development, can be made public by use of the symbols it has created. Unlike perception it is stimulus free, expansive, creative. It thinks, but thinking does not define it, unless the definition of thinking be expanded to beliefs, imagination and will. It looks down at lower levels and constructs concepts from the input. Unlike the perceptual mind that reflects a private world, the conceptual mind creates a private world that can be shared by language and other symbol systems and signs. Or it can be kept private.

Input to mind begins at low levels; at these levels the input cannot be verbalized. At early cortical levels the information is known but is not known that it is known; it cannot be verbalized. It has not reached conceptual mind. It is conscious but not self conscious. It may produce behavior that can be recognized by the performer but the "reason" for the behavior is not known. It is mindless or thoughtless. It is not aware of itself. It is not until conceptual mind is attained, accompanied by language, that awareness of awareness (self consciousness or consciousness of itself) appears. Conceptual mind can be considered the source of an awareness of self—the locus of selfhood.

The composite abstractions of life's experiences, shared and private, make up the self. They are stored in conceptual memory and are coterminous with all conceptual memories for all conceptual memories, real or imagined, comprise the self.

www.ingramcontent.com/pod-product-compliance
Lightning Source LLC
Chambersburg PA
CBHW021046180526
45163CB00005B/2308